James Whittaker's

Little Book

of the Future

DEDICATION

This book is dedicated to the young.
The future belongs to you.

CONTENTS

ACKNOWLEDGMENTS

This book wouldn't be possible without all the
students and colleagues who sat through my classes,
debated over beers and otherwise challenged my
worldview. The future is a big subject and it took
many people listening to my predictions and adding
their own color for me to get this to the point that it
became book-worthy.

I raise my glass to everyone who helped me be
smarter. Insight is the greatest gift of all.

ORIGINS

*In which we go back in time to scrutinize
data's history and puzzle through its future.
And, oh baby! Does it have a future!*

In the beginning, there was data

Data goes all the way back to the beginning of our species. That's right, throughout our long history, we humans have been data collectors and analyzers. Our success as members of our species depended on how good we were at data.

If you think that the current data disruption – you know, the one where machines take all our jobs and render us redundant (more on that later) – is the first-time data rose up to disrupt the world, then you really need to read this book. You see, data has been the cause of nearly every technological innovation the world has ever known, even back to pre-historic times.

And in each one of those disruptions, the people who were slow to adapt faded away to irrelevance. Data – the way we collect, analyze and store it – has ever been the harbinger of huge societal shifts in individual success and collective power. When societal norms around data usage are disrupted, as is happening now, tomorrow's winners are made and yesterday's incumbents are swept aside.

This makes understanding today's data disruption crucial for both individuals and companies who want to stay on the right side of history. The fact that such a disruption is happening *now*, and that it is likely to eclipse all the disruptions that went before it, makes understanding it that much more crucial. Unless you are ready to retire and ride out what's left of your life, you need to understand the future you face.

That's what this book is about. It is an attempt to help wrap our collective heads around what is happening with all the technology that is leading this new disruptive wave: the cloud, artificial intelligence, bots, machine learning and – a little further out – the singularity and quantum computing. It is written for people of all specialties and backgrounds and won't take a degree in computer science to understand. Unless your smartphone confounds your every attempt to use it, you're going to get a lot out of this little book of our near, medium and long-term future.

So, let's get to it. The future waits for no one.

Data and its world-changing impact

Data disruption. It begins, well, at the beginning. The very dawn of our civilization.

You see, our hunter-gatherer ancestors survived

because they were able to turn their observations about the world into actionable data. Weather patterns? Data. Herd migration? Data. The right environment and soil composition where those tasty plants grow? Well, that was data too.

Our ancestors survived by making observations about the world they lived in, converting those observations into data and acting on that data to help them better navigate their world.

Ancestral noggins ill-suited to the challenges that data presented tended to die young, reducing their chances to procreate. They missed meals because game patterns eluded them. But those smart enough to separate pattern from coincidence and convert information into knowledge ate better, lived longer and got busy with mates capable of the same. You and I, well, we both pretty much descended from ancestors who were good at data.

Skill with data: the ultimate evolutionary advantage. And ever has evolution favored those with advantage.

Disruption	Out with ...	In with ...
Spoken language	Brute force	Strategic thinking, communication
Written language	Experience-based observation and action	Attention-based education and analysis
Distribution	Word-of-mouth, local audiences	Data collection, storytelling
Digitization	Closed networks, data siloes, protectionism	Coding, design, user experience, creativity
Mechanization	Humans	Machines

The evolutionary incentive to get better at data has sparked a long series of historical innovations around observing, collecting, storing and distributing data so the advantage could be realized. When something is as important as data, innovation tends to surround it.observing and interpreting data made way to communicating data as data's clear and present value sparked the need to convey it to others so we could pass along that value.

The language disruption gradually increased our communication skills and enriched the kinds of conversations our ancestors could have. Through language, we could pass along the wisdom of our collective lore to our children and collaborate with our tribemates. We could plan hunts and work better together to conduct warfare and secure a food supply. The ability to learn language, and get good at it, helped our ancestors thrive. It gave them a leg-up over competitors who weren't as good as they were.

That's what disruption does. It changes the game. Individual abilities around data interpretation were suddenly less important than the ability to communicate that knowledge and get others working together as a team. Great communicators suddenly found their skillset highly valuable. The better your language skills, the easier it was to get by and to thrive as a valuable member of your tribe.

Language, however, was only the first data disruption. The next disruption came in the form of codifying language into written form: the dawn of recorded history.

Talk about a disruption! Once data, lore and tribal wisdom was committed to writing, it could survive its owner. The world no longer needed to be experienced first- or even second-hand. It could be read, studied and analyzed like a science experiment, requiring new analytical skills and allowing an emerging segment of society to succeed in ways that weren't possible before. You see, each disruption elevates certain skills and deprecates others. The new abilities to read and to write created advantages for those who cultivated them. The attention span to study, infer and combine old knowledge with new insights created advantage for those who could do it. Those skills had never before been required and those who nurtured them rose to the top.

The ability to write created power like the world had never seen. It was a rare skill and those who commanded it could not only gain employment where others could not, they could determine what information others were exposed to, influencing future generations in the most profound way imaginable.

You see the pattern, right? Disruption changes the game. It moves the cheese. It shifts the balance of power. It creates advantage for some and removes advantage from others. When people and organizations are slow to adapt to the new disruptive technology, they are left behind.

I hope you are seeing the purpose of this history lesson: getting ahead of disruption grants you advantage. Dallying in the old ways is a great way to ensure you have a poorer future. It pays to focus on *what's next*.

Still disruption was not yet done. Whenever there is advantage to be squeezed out of data, you can guarantee people will innovate to do just that.

Not long after data began to be captured in written form, the need to collect and protect it arose. Enter the great libraries. Build one and sure enough, a mighty civilization would rise to keep it company. Institutes of higher learning came hand-in-hand with the prevalence of written knowledge. These learned institutions then created even more data and pushed the world into a new era of education and enlightenment. The scholar class was born and assumed its place advising royalty and guiding the world into the future. Education based on the written word remains an advantage to this day. (You are, even now, seeking that advantage by reading this book, aren't you?)

Once again, skill with data created advantage to those who mastered it. Those who lacked access to such data, or failed to embrace it, were left behind and became economically and culturally disadvantaged. It's no coincidence that most of the civilizations that built great armies and conquered the world had great libraries and universities. Shared knowledge creates tradition and culture; both of which are worth fighting for.

The mass production of data via the invention of the printing press was the next data innovation. Its appearance disrupted the status quo in a major way. Data, and the knowledge extracted from it, could now be printed and distributed at scale. Radio and TV, centuries later, would only grease those distribution skids and increase the reach and value of those who were good at data.

During this long phase of data disruption, it was possession and distribution that brought the most advantage. The companies that built data repositories (literally warehouses full of file cabinets which, in turn, were full of paper and microfiche) and acquired an audience around its distribution were the new winners.

This was the dawn of the media companies and they ruled the data world for decades because they possessed, collated, analyzed and disseminated data at scale (and with commensurate profit margins).

It was a hard business model to replicate (as it required vast investments in storage space and copies of the data that filled it). This limited the number of winners. Global, and even local, news agencies were few and far between because data was so siloed and hard to replicate.

Media companies used their power to, quite literally, create culture and were star makers in music, film and business. None more powerful than the likes of MTV in the 80s and 90s. They were kingmakers and curators of culture for an entire generation. All because they mastered the skills and business models of the current disruption.

Data is power. World-changing power.

But data still wasn't finished evolving and the next innovation, the digitization of data, would make possession of data moot as it could, finally, be easily and precisely copied. At the same time, computer networks would rewrite the rules of data distribution, disrupting newspapers, radio, television and content industries like music and movies. Out with the incumbents, in with the upstarts. Each disruptive period has its winners and losers.

Think about it, who's going buy an album when they can download a perfect copy of it for free? Old media didn't see it coming. They didn't react to it when it did come. Exit the power of newspapers and radio – and even the mighty MTV – and enter the tech giants.

The pattern keeps repeating, doesn't it? Out with the old and in with the new. Once again, we have the forward-thinkers disrupting the winners of the prior epoch. When it comes to data, no one is above disruption. No one can afford to be complacent. Incumbency is one giant ball-and-chain.

Online news organizations taught this lesson to the established networks. Napster taught it to the music labels. Netflix taught it first to the video stores and later to Hollywood. The old winners became the new losers.

Thus, has it always been since time immemorial. The only thing constant and predictable about the world of data is its accelerating rate of change.

The digitization of data ultimately led to modern computing. Computers, the web and mobile devices are all testament to the value of data and our need to process it at scale and access it no matter where we are day or night.

Think about it. Do you possess the machines in your pocket and your backpack because they are fashion statements? Because you like the way they feel in your hand and on your lap? Because the marketing was so slick you just couldn't resist? No. You possess them for the access they give you to the data you want and need in your life.

Data is now so important we are addicted to it. The idea of living a data-free life is unthinkable and would put us at a huge disadvantage in both living and working.

Clearly, the new advantage in this data-addicted world lies with the software professionals and programmers who know how to wield it. Those who understand the arcane languages of computers – and understand how data flows around the internet – have access to career success that those who lack these skills do not. Coding is the new reading, writing and arithmetic combined. The richest, most recognizable people in the world are no longer entertainers and heads of state: they are the ones who lead data-intensive companies: Steve Jobs, Elon Musk, Bill Gates, Mark Zuckerberg and their like. Compare their net worth to old media and you get a sense of the power of leading a disruptive wave. But they are just the latest in a long line of post-disruption winners. Ever has data been the cause of innovation and the source of wealth and success. Never more so than now.

But even in the time of the world wide web, innovation around data is not yet done. This next innovation, the one happening right now, is the subject of this book. This new era will bring disruption and change the success landscape once again as it shifts advantage away from human intelligence and ingenuity and, for the first time in history, gives that advantage to a brand-new species: the intelligent machine.

That's right, our ability to keep pace with data has ended. All the advantages of this new era will lie with machines *because they are far better at data than we are.*

Will those machines remain tools that we wield at our pleasure? Will they stay subservient to our will and act as personal assistants helping us navigate the data in our world? Or will they work alongside us as colleagues? Perhaps, they will become as intelligent as we are and develop skills and occupations that are as valuable as anything we humans are capable of. They could even become so indispensable as assistants and companions that we can no longer live our increasingly complex data-driven lives without them.

Or, perhaps, as many are now warning, they will prove to be a separate and superior species – because they are better than us at data – and thereby render us redundant as every Master of Data has done before. If history does, once again, repeat itself then the human species is in serious trouble.

Make no mistake, we shall be the first generation of humans who will share the world with a species smarter than ourselves: artificially intelligent machines that exceed our own ability to navigate the world around us. When, one day, the machines grow smart enough that they no longer require us to build or maintain them, what destiny will they choose for our own species? When we are no longer necessary to their survival, what then? When the advantage shifts to the machines will they remain our servants, become our colleagues or take their place as our masters?

Your future is data, get ready

Given the advantages that data will serve up to the machines, the future doesn't look particularly bright for humans. In this final disruption, the new losers are humanity *in its entirety* and the new winners are our machine assistants/companions/nannies/overlords.

Our generation is the first generation smart enough to build machines that are our intellectual superiors and yet dumb enough to actually build them.

But the time of the machines has not yet arrived and this final data disruption is just now gaining momentum. We humans are still in control of our destiny, for now, and the decisions about how to build, train and program these machines are yet before us.

This means that we need to get good at data. The more of us who understand data and the potential of the disruptive innovation that is currently playing out, the better chance we'll have of making informed decisions about how to handle the machines that we will inevitably build. The more data-educated voices we have to debate the possible outcomes, the better those outcomes might be.

That's what this book is about. Your pint-sized but knowledge-packed guide to the world of machines that is coming for your jobs and threatening your future far faster than you think.

~ 2 ~
RISE OF THE MACHINES

In which we catalog the various technologies
involved in the coming machine revolution.

The cloud

At the center of the current data disruption is a technology called *the cloud*, and it has transformed the storage and organization of data. That data has, in turn, enabled algorithms to begin getting really smart. Without data, the algorithms have nothing to work with. The cloud, and its ability to gather all the data in the same place, was the game changer.

More about those smarts in a bit, but first a little explanation of what the cloud means is going to help us understand how these algorithms work and speak to their ultimate potential.

Before the cloud, data was distributed across an entire world (literally the world wide web) of individual web servers. These lonely servers knew little about each other and the world around them. They stored data and they served data. Siloed islands that contained only nuggets of knowledge.

Boring, right?

Algorithms, no matter how clever, could do little more than look at the data on a single siloed server. They couldn't process and learn from data across servers because that was too hard a problem given the storage and network speed constraints of the time. Algorithms couldn't connect the dots that a contiguous world of data might have allowed them to, because they could only access data within a single silo. Combining all that data into a more useful whole just wasn't practical.

The cloud made it practical. As data migrated from geographically distant servers to being co-located in data centers, the world's information sat together for the first time in a convenient place for algorithms to access it *as a whole*.

And that changed everything.

This co-location of data is important. Crucially important. The cloud took those data nuggets and combined them into a mountain of knowledge. Patterns and answers that weren't discernable before suddenly were, because *all* the pieces fell into place. The algorithms finally had the data they needed to start becoming smarter.

All this data sitting in one place has resurrected an idea that dates from the late 1950s: that computers can learn. How? From data, of course. Give them enough data about a subject and they can learn that subject, or at least synthesize patterns from it.

Once a big, fat theoretical *maybe* from optimistic

mathematicians and computer scientists, the renaissance of algorithms came to light within the cloud. With enough data, algorithms might just find patterns to create something approaching a thinking, intelligent machine.

Two ideas central to the possibility of machine intelligence are what we turn to next.

Machine learning

The first idea is *Machine Learning* or ML for short. The concept is simple: instead of programming a computer the ordinary way with exact instructions that operate over a fixed input domain, an ML algorithm is given a large set of data and "trained" to find patterns in it. Or, more simply, traditional algorithms are told exactly what to do. ML algorithms are given data and asked to figure it out themselves. This means looking through all the data and finding patterns. Indeed, way back in the 80s and 90s this field was called 'pattern analysis' and 'pattern recognition' long before it became machine learning.

For example, take a bunch of pictures of cats and inform the ML algorithm that *these are cats*. The ML algorithm will then analyze each one and try to figure out what makes them cats.

Now, this happens at a technical level, not the intuitive way it does for us humans. An algorithm needs to turn 'cat' into data. In other words, it needs to find out what each of the pictures has in common at the pixel level so it knows, with data, what this 'cat' thing is. Maybe it's the distance between the eye and the tip of the ear. Maybe it's the shape of the nose or the geometry of the outline of the head. Or, more likely, it is a combination of several observations that allows it to classify an image as 'cat'.

Now, take a bunch of pictures that don't have cats and inform the algorithm of this. It will again reduce the photos to data and try to determine the math of 'not cat'.

Given a large enough dataset, ML is pretty good at then taking a completely unseen picture and figuring out whether there is a cat in it.

You can see that the larger and more diverse the "training set" of cat pictures, the more patterns of cats the ML algorithm will recognize. That's where the vast storage of the cloud comes into play big time. If you have a large enough collection of cat photos, the cloud is a perfect place to store them. The cloud is a one-stop shop to train your ML algorithm.

Now the more data an ML algorithm has access to, the better it can get. Just like humans studying, the more information and the longer the concentration, the more we can learn.

But that's where the similarity stops. ML learns either by sheer brute force, shifting though vast amounts of data looking for patterns (this is called unsupervised learning) or by being told, by a human, certain categorical truths that help speed it along on its task (which is called supervised learning).

For example, we can show an ML algorithm a shit-ton of cat photos and, without telling it anything, it eventually notices that each of the photos share a common characteristic of a cat shaped object in them. Or, imagine an algorithm that watches thousands of hours of chess matches and can categorize bad moves and good moves out of sheer volume and brute force. That's unsupervised learning. It works. Sure, it takes more time, but machines have nothing but time. Still, if you want machines to learn even faster, well that is possible with supervised learning.

Supervised learning shortcuts the time it takes machines to learn. We might tell the ML algorithm that a cat has ears, which might make the process of identifying them easier. Or, we might tell the ML that the queen is more valuable than a pawn or that the king must be protected at all costs. In each case, the additional information helps the ML algorithm learn faster.

Of course, there are issues of data quality and outliers, like un-catlike felines and catlike nonfelines that will confuse the ML algorithm. Indeed, the algorithm will even be wrong at times (just like us humans can), but the more data and the more time the ML is given to learn and refine, the better it gets.

Machine learning is a long game and machines themselves, well, they don't mind. They are tireless and revel in the long game much more so than humans ever could. After all, how many of us take the time to get really good at chess?

ML algorithms have already amassed some serious accomplishments. For example, they learned how to spell and do it far better than we do. They can also recognize bad grammar, having been trained on millions of documents that show examples of both bad and good sentence structure.

That's right, those Microsoft Word features you've been using for decades had their humble beginnings as ML algorithms. They've gotten so good at it that they are better than all but the most expert humans.

Machines have also learned to play chess and Jeopardy and Go. The latter was, at one point, considered too strategic for a machine to ever comprehend. But that was before a machine beat, soundly, the best human Go player on the planet.

You can begin to appreciate that, eventually, machines will get good at things like recognizing the face of a known terrorist at a rental car counter. Or identifying a sex offender prowling a mall or approaching a school. Or recognizing invasive weeds in a field of edible crops. Or identifying cancer cells among a high-res image from an MRI scan. Or finding new chemical combinations to create better medicines.

Machine learning has much to offer humanity and even now millions of machines are learning from exabytes (that's 1000^6 bytes, or, to get technical, a shitton) worth of data about subjects from art to health to finding habitable planets outside our solar system.

As you might expect, with all the data the cloud has to offer and nothing but time to process it, the machines are getting good at a lot of important stuff. The question is, where does this learning stop? Having first, for example, mastered spelling and then grammar, will they learn to write next? And if so, will they be able to write better fiction than us, report more consumable news or craft a better legal argument? On what sort of writing tasks will humans maintain their current advantage and where will machine-generated content reign supreme? What about making movies? Music? Art?

It is hard to find a place where this learning will stop. Will we soon be asking existential questions about what we know as humans that won't eventually be learned by machines? What part of the human condition cannot be reduced to data so that the machines will learn to do better? When machines have learned to navigate the world without our help, what will be left for us?

And, fortunately or unfortunately (yes, the jury is still out about whether ML is a good or bad thing for humanity), ML is not where it stops. There is another branch of computing that is even scarier.

Artificial intelligence

Enter the second type of algorithm disrupting our future: *Artificial Intelligence* or AI for short.

Some in the industry conflate AI and ML and it might seem like splitting hairs, but they are different in both form and function. Form: Where ML uses math and statistics to look for patterns, AI attempts to simulate the way the brain itself perceives the world using a branch of mathematics called *stochastic processes*. Function: Where ML is concerned with classification (cat or not-cat), AI attempts to problem solve and take action. But they are inextricably linked. Learning begets intelligence; no one gains intelligence without a great deal of learning. This makes ML and AI the perfect allies in the pursuit of data insights.

Let's pick this apart. The differences may seem trivial to you, but the thought processes that go into AI as opposed to ML are fundamentally different and often lead to different solutions.

Given that the human brain represents *natural intelligence*, copying its structure with "artificial" programming constructs gives AI its interesting (and rather intimidating) name *artificial intelligence*. In general, an AI algorithm must mimic the way a human brain works to be called AI.

Scientists fell on this line of reasoning as early as the

1950s based on the observation that even simple clusters of neurons can perform very complex behaviors. If these neural clusters can be simulated with programming constructs, machines might be created that "think" the same way humans do.

Fascinating, right?

And also insightful. If we want to emulate the structure of human thinking with machine thinking, we need to code things that work like the human brain.

Imagine those prehistoric ancestors of ours observing herd migration. Those observations cause brain neurons to activate. Given a specific sequence of observations (say, the game pauses at some watering hole), a set of connections between the neurons happens. Our ancestor observes that there is a connection between the herd and the water. Learning occurs. As they make more observations over some period of time, those neural connections are reinforced and knowledge about migration patterns emerges and becomes useful.

This is how humans go from data to structure to knowledge.

To do this with machines, we need programming constructs that perform the same function as the brain components involved in learning. So, individual neurons in a human brain are cast as individual elements of data storage. *Data structures*, we call them, that store specific pieces of data. Pathways connecting individual neurons become datum organized, connected and ordered via links. A link between two data might imply order. Another link might imply association, another will represent relation, another will describe subsumption and so forth. Instead of neural pathways being reinforced based on observed behavior, links between datum are given higher or lower numerical weight based on their context, just like neural connections are reinforced in our own brains. Get it?

Look, don't sweat the details, just understand that every mechanism that we understand in the human brain has a counterpart in the AI code. The brain itself is modeled with constructs related to data. The operation of the brain is simulated by noting the order and frequency of those datum. These structures often take the form of matrices, nested matrices, state machines and stochastic processes. Viola, we just modeled the function of the brain with well-studied branches of computer science and mathematics.

Aren't we clever.

That's right, AI is nothing more than higher math encoded into a computer program. The magic of math made physical with the magic of computers.

For example, IBM's *Jeopardy!* playing computer Watson takes a layered approach to answering a quiz question. First, it ingests dictionaries, encyclopedias and documents with relevant answers. These become nodes in its "neural network." Second, it applies 100s of different language analysis algorithms to extract meaning from questions and uses that analysis to connect various nodes that have related information. Third, it reinforces connections based on the consensus of those algorithms. Finally, it executes code to check if the consensus answer makes sense given the context of the question. So, when Alex Trebek asks it a question, it follows the neural pathway that most looks like the question, leading it directly to an answer.

The more it plays, the better it gets as relationships between question, structure and answer build up over time. At some point it is going to be so good a human will never beat it. After all, a human player has a life outside the game, a machine can keep learning 24/7.

Assembling programs in the same manner that a brain builds neural connections has made AI algorithms incredibly effective at certain classes of problems, like speech recognition (we learn to recognize voices easily) and game playing (we get better rapidly as we play). AI thus represents a potential upgrade over ML. For example, machines can learn the letters and words in a language like English, but AI can interpret the words and extract meaning from them, even translate them into other languages.

Where ML finds patterns, AI finds insight.

If that seems too subtle for you, don't worry about it. They are just words. It's the general concepts that matters. Machines aren't just observing our world and learning from it (ML), they are going to be able to act on what they know (AI).

AI has some fancy sounding structures like *neural networks*, in which networks of small data connections form as new information is learned by the algorithm. As new connections are made based on learned relationships between datum, the algorithm can infer new knowledge and make new connections.

Artificial constructs (that represent the neurons in the human brain) connecting to other artificial neurons based on some observed relationship is what AI is all about. The more we learn about the brain, the more we can mimic it with programming constructs that simulate the same kind of learning processes. As these artificial neural connections increase over time, the neural network becomes smarter and able to solve increasingly harder problems. Just like us humans as we

21

grow from infancy into adulthood. That's right, over time AI will get smarter just like we do. But given AIs attention span, tirelessness and lack of anything else to do, it is going to get a lot smarter, a lot faster.

A specific AI technique you might hear about, especially lately, is called *deep learning* and it takes the neural network one step further by mimicking the way humans learn: in stages using specialist neurons. One layer of our brain makes observations and passes those to another to make insight. Those insights are passed to yet another layer for decision making and taking action. The layers and sections of our brains all operate differently and specialize on narrowly-defined tasks. The magic isn't in any specific layer, it is in the collaboration of the layers and the operation of the whole.

Deep learning AI simulates this process, except the layers (often neural networks themselves) are artificial programming constructs. The layers collaborate (propagate is the term AI scientists use) by sharing information among the various layers to affirm or change the knowledge the network possesses. This way, over time, the network becomes more and more capable as it learns what information is useful and what is not and the context of both.

But don't worry if all of this sounds arcane, the point is that programming is not what it used to be (take an input, add it to some data structure and then produce an output, rinse-and-repeat). Traditional software does the same thing every time it is executed. AI does not. A single new input can have cascading effects that changes the way the entire network behaves.

When we program AI, we code differently. AI code

can react to its environment. It is capable of a different reaction given the same set of inputs based on changes in context and history of operation. AI matures and grows where traditional software does not.

Sounds like science fiction, doesn't it? Well it's not. AI is making its way into real products that solve real problems:

- IBM built an AI that can play chess and another that can play Jeopardy. Both are at or above the capabilities of expert humans.

- Skype has an AI that is capable of real-time language translation, allowing two humans to seamlessly communicate even though they do not know each other's language.

- Google built an AI that plays Go, a strategy game once considered out of the league of AI and, again, at or above expert human levels of competence.

- Microsoft Kinect (and its successor HoloLens) can watch your movements and render a 3-D avatar of you that it can then interact with. This is an exciting development because computers are now entering our world and allowing us to enter theirs. Thanks to AI, the boundary between the cyber and the physical world is beginning to blur. (More about this topic later in this book when we get to holograms and the idea of augmented reality.)

Of course, since the data and algorithms that power these networks aren't static, the more they play, translate or watch, the better they get at playing, translating and watching. One can imagine over time that AI will make better doctors, lawyers, bankers and just about everything else as they ingest new knowledge and tirelessly toil to make sense of it.

These and many other AIs are performing at human (as good as we are) and even superhuman (better than we are) levels. Many more are getting there. Remember when I talked about all that data in the cloud and all those tireless machines processing it? Well, in the time it took you to read the paragraphs that got you from that section to here, a great number of those AIs got a whole lot smarter and another large set of AIs just began their long study. They aren't sleeping, they aren't eating, they will never stop learning.

And, now that most of the things people do are in digital form from inception, machines have easy access to the detailed data of our everyday lives. It would be reasonable to assume that the pace that machines learn will quicken and the capabilities of AI will grow. As more people become capable of programming AI, those AIs are gaining access to more data than ever. They are learning about us. They are learning about the world. They are learning about everything.

So, who will benefit from this new, modern world of data and intelligent algorithms? Clearly, it will be the people and organizations who have the infrastructure to amass digital data (think data center operators like Microsoft, Amazon, Google, Facebook and Chinese companies Baidu, Alibaba and Tencent) and the talent to convert that data to useful knowledge (think all the above again). Plus, there are the companies whose machines and apps are in the hands of users collecting their intent at the spot it occurs, which is some of the most valuable data on the planet (add Apple, Samsung and even Twitter to the above). These companies will continue to outpace more traditional software companies who are going to have to figure out their own AI story or join the likes of once-important companies like DEC and Sun Microsystems in the technology cemetery.

The successful will be the ones that identify and collect meaningful data sets and possess the skills to train the models and build the algorithms that will learn from them. Tax companies need to view tax data differently. Insurance companies need to view insurance data differently. Banks need to view financial data differently. Hotels need to ... well, you get the point.

The world of software is giving way to the world of data, models and algorithms and the hardware (machines) that embody those algorithms.

Don't believe me? Compare the net worth of companies like Amazon, Apple, Google, Facebook and Microsoft to the top companies in any other industry. You can't because it doesn't. Compare the wealth and fame of Silicon Valley entrepreneurs and venture capitalists with their counterparts in Hollywood or on the Hill. Traditional wealth centers pale in comparison to tech.

The writing is clearly on the wall. The balance of power has shifted toward the kings and queens of data. It has ever been that way and will likely stay that way in perpetuity.

If you aren't already looking into getting good at this new world of digital data, you might want to start. Otherwise, you could end up like our ancestors who never could get ahead of herd migration. You know, extinct.

~ 3 ~
BIG BETS

In which we discuss projects really smart
people are working on and the kind of future
those projects are creating right now.

No data like big data

Now what to do with all this data and AI? That seems to be the big question for most individuals and companies trying to succeed in the coming years.

Well, if you haven't spent a lot of time thinking about data, ML and AI then you are at the right place. Because I have.

You're welcome.

And any discourse on this subject should start with the big bets that top technology companies, Silicon Valley backed startups and the most innovative researchers are placing. These are the people and organizations producing the moonshots that will either pave the way to the future (assuming they are successful) or inspire others with their audacity. We could do a whole lot worse than following their lead, or at the minimum, learning from their example.

Here we go:

Video and sensor surveillance represents the first such moonshot. The hardware (think Microsoft Kinect) that allows computers to be aware of their environment has undergone serious technological innovation since the Xbox and much of it is ready for primetime. All it needs are a bunch of smart developers building the right set of scenarios for it. Consider:

Cameras on both Windows 10 and iOS are already good enough to distinguish their owner's faces and prevent access by other persons. But the technology goes far beyond facial recognition. Buildings, city blocks, and soon entire towns, are being made smart by a network of cameras watching traffic flow, foot traffic and even activity inside stores and businesses. Schools will be made safer, shoplifting will be prevented and criminals will find fewer and fewer places to hide.

Soon, *the city itself* will be able to guide you to the nearest parking garage or coffee shop. All you'll have to do is declare your intent to find a happy hour and businesses will immediately compete for your custom.

Stick around, it gets better.

The internet of things

Loosely, this kind of technology is called *the internet of things* or IoT for short. Think about it this way: whereas the internet we all know and (mostly) love is a network of *computers*, IoT is a network of *common things*.

That's the difference. Imagine an internet of ovens which all know how to recognize food and cook it in various ways. An intelligent oven will make anyone an expert chef. Just add food ingredients and the oven does the rest.

Those ovens will talk to each other across town and the world to share recipes and advice. One oven learns how not to burn a pizza and all ovens instantly share this lore. ML and AI will ensure that the collective knowledge of the internet-of-ovens knows what all food looks like and how to cook it. They will even be able to prevent food-born illness. Their ultimate task to better serve their humans will require that they recognize, learn and communicate so they keep getting smarter.

IoT scales up from individual devices to collections of devices that collaborate to serve humans. The internet of ovens scales up to an entire smart house. Smart houses scale to a smart neighborhood and eventually to a smart city.

Imagine a local bar and grill with sensors under the table to recognize when the table is free or busy (bent knee recognition?). Perhaps it learns the pattern of a busser shuffling around cleaning its top and figures out which tables are ready for new customers. The moment a table is available to new customers, the restaurant (or rather the machines that automates it) knows.

This bar and grill will be able to learn how long it takes for people to be served. It might recognize gestures made by customers that indicate they need a waiter. It can learn which dishes are popular at which times and assist in the ordering of ingredients. It can keep learning for years, improving operations, service and customer satisfaction. It can communicate with other restaurants making the collective whole even better.

Before too long, no one will want to dine at a human-managed place except, perhaps, for a bit of nostalgia. You know, just to see how grandma and grandpa used to live.

A customer should be able to walk up to a smart eatery and ask the building itself, "is it happy hour and is outside seating for four available right now?"

Of course, the specials, soup of the day, rotating taps and wine lists will be at the ready for such an AI. It never forgets, is never bothered by a bad mood or troubled by personal problems. And it is never too busy to help because it can do all of what it does *at speed*.

The restaurant knows it's a restaurant.

When you look at the world through an IoT lens, the possibilities are endless. Suddenly that age old "what do you want to do tonight, honey?" question can be answered by the bars, venues and shops that offer 'things to do.'

How much better is this than Googling for an answer and sorting through a sea of irrelevant links and scores of ads? Much better. You see, in a world of IoT, you aren't responsible for finding 'something to do' all by yourself. The devices that represent all those 'something-to-do' establishments will be looking for you too.

Personal agents

Here's where the big bet of personal agents, or personal AI (which seems the more accurate term) surfaces. If a machine (maybe you wear it, carry it or have it watch over you from above) has learned your likes and dislikes, favorites and pet peeves, schedule and habits, moods and broods, etc., in the same manner as it learns how to pick a cat out of an image, it will know what you need and can negotiate with the relevant IoT devices on your behalf.

For example, if you declare your intent to dine, your own AI will know what kind of food you like and whether you are in the mood to drive a long distance to get it. It will communicate with the various restaurant AIs and find out which have tables free and negotiate a transaction.

This is good for the restaurants too. On a busy night, they can avoid frustrating customers who might otherwise show up and have a poor experience. On a slow night, they can be aggressive about negotiating for your business.

This is a fundamentally different kind of match making than browser-based search. When you Google for something cool to do in your neighborhood, you are the only one participating in this effort. The web is static. The information about what you might do is sitting there motionless waiting for you to find it.

On the other hand, the IoT and AI model is *bi-directional*. You are looking for something-to-do and that triggers effort by something-to-do, simultaneously, to search for you. You are looking for a restaurant and the restaurants are looking for you. Finding a place to eat is no longer your problem; it takes a village.

You won't remain ignorant of new restaurants in this future either. That new restaurant will find you. Similarly, you won't miss your favorite bands coming to town. You won't miss the best sunset, happy hour and date night. You might even stop asking "what do you want to do tonight, honey" because the answer is already waiting for you thanks to the AIs planning, watching, analyzing and cataloging what people are doing and how much fun they are having doing it.

If these walls could talk

Innovation in the IoT area is going to occur in the workplace too. Conference rooms will understand their schedule, what meetings are taking place in them, who is in those meetings and when the room is free. Company cafes will place your order and direct you to the table where and when it will appear. They will know if you are in a hurry and expedite your order or have a small robot deliver it hot and fresh to your next meeting.

Oh, there is so much more, keep reading!

I like to think about the internet of things in the following way:

If this thing *could talk, what would* it *say?*

This question leads to a fruitful discussion about the potential of any specific device, appliance or building with respect to its activity on the internet of things.

For example, if a conference room could talk, what would it say?

With that as the guiding question, it's time to brainstorm:

First, a conference room would understand its schedule and know who was attending each meeting. Seems simple enough, right? That data is, after all, readily available in employee calendar apps.

Collectively, the conference rooms in any building

would know each other's schedule and guide any employees with the intent to meet to a free room suitable to their group size, purpose and time requirements.

One could ask a conference room, "is everyone here?" The answer is discoverable based on the meeting invitation and recognizing the faces of the people in the room. Furthermore, a missing person might be geolocated using, say, their phone (they are, after all, employees of the company) and an ETA could be established.

There are a lot of answers that once were the province of a human admin or receptionist that will be handled far more efficiently by machines.

The machines, however, can go far beyond what humans could do. What if, for example, the parking garage under the building was also on the internet of things? The conference room might inform the garage to reserve a parking spot closest to the elevator bank that will get the wayward meeting attendee to the room as quickly as possible.

This is where the 'internet' in the 'internet of things' comes in: when each self-aware/task-aware machine communicates to other self-aware/task-aware machines to get work done, well, that's where the magic happens. The machines are much more insightful and useful as an ensemble.

See what a simple question like 'what would this machine say?' sparks? We just designed a smart building together. Look how fast you are learning!

There are no limits

In a world where every device is aware of its purpose and its human users and can self-maintain and repair, the number of things humans are needed for begins to get small. As the world is inexorably reduced to data, the machines will take over aspects of our lives that used to be the sole province of people.

Let's turn to some more big bets now in other forms. Even the sky isn't the limit.

~ 4 ~
THE WORLD REDUCED TO DATA

*In which we show how to reduce problems to
data and argue that nearly every problem
reduces to data.*

The age of machines

When you look at the world as data, it isn't hard to imagine that machines are going to get good at navigating the world *without the help of humans.* Restaurants that order their own ingredients and find their own customers have little need of management, advertisers and marketers. The only help they will need is in the data science bit.

That is until they learn how to do that too.

You see, the machines will approach problems a lot differently than we do. When we look for restaurants we are looking for *possibilities.* When AI looks for restaurants it is looking at *data.* What restaurants are local people really enjoying right now? Which ones have no wait? Which ones serve food that my human likes? Which ones have the ambience my human wants tonight? What is the expected cost and is it in line with the norms for my human's spending patterns? My humans want to be alone to talk, which restaurants aren't frequented by people they know so they can be steered clear of those?

This is how the world of data looks to the machines.

These questions would take a human with a search-engine hours to explore. A machine knows the answer instantly.

And a lot of questions fall into this area of machine expertise, for example:

"How much would it cost to take my family to Europe for Christmas?"

Go on, fire up your browser and come up with an answer, I dare you. You'd have to check every city in Europe one airline and one flight at a time.

But machines? Well, all the families from your hometown who have gone to Europe during Christmas season is discoverable. It just takes the right organization of the data and a powerful algorithm to rip through it looking for the answer. Believe me, the data is there, and someone is going to build that AI and start of a new genre of travel-dominating companies.

The travel industry is rife with such questions: what hotels do people like me usually pick in San Francisco? Do people usually rent cars or take public transit when visiting Denver? What is the biggest music festival in greater Toronto during the summer?

The machines are going to be especially good at travel because – like law, finance, accounting and politics – it's already mostly data.

Pick a question, any question

Almost any question you might ask has data to help answer it:

"My energy levels have been really low since I turned 50." Go on, Google it. You're going to get snake oil, homeopathic bullshit and nonsense lifestyle recommendations as far as the eye can see. Ads are a terrible discoverability mechanism. They don't give you the best answer; they give you the answer the advertiser wants you to see. Think Google is on your side? No, Google is giving you the answer it has been paid the most to give you. An ad is, by definition, the front-end to someone trying to sell you something whether you need it or not.

But the data that you need is certainly there. You are not the only man to hit the change of life or the only woman to go through menopause. There are solutions and many of the people who found them share your body chemistry. So, in the quest to discover your own personal fountain of youth, who's going to find it first? You and your browser? Nope, I'm betting on the machines.

You see, if you do manage to find your personal fountain of youth, you have no way to inform the web, "hey, this solution worked for me." But in the bi-directional world of the internet of things, your devices, sensors and monitors will know that you are feeling more energetic and will be able to anonymously report that result back to the data infrastructure. Now, as more people turn 50 and seek the solution you found, their task just got a lot easier.

Machines are that much better at data than we are. The more we cast our needs and accomplishments as data, the better the machines will get at serving those needs and creating more accomplishments for us.

The things machines are going to be better at is a very long list. To wit:

"Let's binge a show tonight!" The expression *800 channels and there's nothing on* is an expression for a reason. Finding something to watch is a time killer. And not because of the lack of good shows, but because of the undiscoverable mess that is the content landscape.

Netflix knows what you like, but only on Netflix. The History Channel knows what you like, but only on the History Channel. HBO knows what you like, but only on HBO.

Not helpful.

But an AI inside your TV would be able to know both who is watching (via voice recognition or camera), what they are watching and learn how much they are enjoying it. It will know what you watch alone and what you watch with your spouse and what you watch with your family. Viewing patterns can be (anonymously) shared across homes so that a couple that enjoyed Game of Thrones on HBO and Vikings on the History Channel might be directed to Marco Polo on Netflix because the pattern of people enjoying that collection of shows is learnable by some ML algorithm that is part of the internet of televisions. Machine learning wouldn't even break a sweat and your experience would be way better than what you have today.

Here's the key takeaway: No matter what you are trying to do right now, from researching therapies and cures to figuring out how to entertain your toddlers on a long drive ... someone has already figured it out ...

likely a great many someones. And as those solutions are gathered and organized by machines, and algorithms are trained on them, the machines are going to find insights and answers for us and provide those answers before we even have to ask.

How many humans will want to return to a world where all the search effort accrues to them and none to the network? The machines are going to help us be healthier, happier and, most importantly, give us time to be humans.

Privacy concerns

Yeah, I know what you are thinking. This is creepy, right? Machines digging into our lives and learning our likes and dislikes.

Is it though?

Most of us have already given up our privacy to Google and Facebook and for what? Better ads?

Wait, better ads? That's it? We gave up all our personal information and in return we get *better ads*? We don't even like ads so the idea of a "better" ad is quite the oxymoron for most of us and it's a poor deal for all of us.

But machines offer us far more value for knowing all those things about us. In exchange for allowing them to know our medical information, we're going to get better healthcare and advice. In exchange for information about our viewing habits, we are going to get a better entertainment experience. We are going to same time, have more intense experiences and not have to stare at a screen all day.

The machines are going to go way beyond "better ads" and using what they know about us are going to make our lives better. They are going to save us time. They are going to bring experiences to us that we might not discover otherwise. They are going to make our lives easier and enrich the experiences we have with other humans. Privacy will cease to be a concern when this value proposition becomes tilted in our favor. The moment the machines provide things we highly value, we will stop talking about privacy. Insistence on privacy will mean the machines won't have enough data to serve you. Your need for privacy is going to make your life a lot harder.

Choosing privacy will mean forgoing the value that machines can give. When the value is clear and present, few humans will choose to keep their data and identity private.

Security concerns

Whereas privacy is a value-proposition, security is more existential. The world will always want data and transaction security. And, here, the internet of things and AI offer some interesting hope.

Mind you, security is, and mostly always will be, an arms race. The good guys invent a new security measure and the bad guys figure out how to circumvent it. The good guys secure a website and the bad guys figure out how to hack into it. The good guys invent crypto algorithms and the bad guys decode them.

This time, however, AI will be watching. It will learn how a secure server behaves and how a hacked server behaves and know how to act whenever the latter happens. It will learn what network patterns are normal and what patterns are attacks and know how to react. When AI is watching, security will be much harder to breach.

But the easiest hacks are never hacks against machines, they are hacks against people and here is where even more hope lies: humans, the weakest security link in any network, are mostly removed from the internet of things. With humans out of the equation, machines are going to be orders of magnitude harder to hack.

Machines don't reuse passwords. They don't need to write them on sticky notes. Machines don't fall for phishing scams or click on dangerous links. Machines know how to authenticate another machine's identity and secure data they communicate to each other. They can do this at scale and at speed. Without humans in the loop, we have a chance at real security.

But like any technology, we must take security seriously and we don't have a good track record of doing that for any technology, like IoT, in its infancy. As an industry we didn't consider securing our PCs until they got hacked. We didn't build a security model for websites until they got hacked. We invented technology like cross-site communication without realizing its potential to get hacked with cross-site scripting. Security has always lagged new functionality.

We cannot afford to do this with IoT. The machines are representing our identity, our data and our purchasing power. Security is more important than ever.

What an opportunity for both your career and your company to take security seriously. Fortunately, we have the poor example of PCs and the web to learn from. This is a track record we do not want repeated.

This time around, we need to get security right ... right from the start. And with humans out of the loop, we have a good shot at doing just that.

And then there is quantum computing: a chance at nearly unhackable machines. But that is a subject for later in this book.

What doesn't reduce to data?

Some fields have already been completely reduced to data. Driving a car, flying an airplane and operating most machinery. Robots are getting good at bussing tables, weeding our lawns and cleaning our houses. They even make good repairmen. Workers in these fields have a bleak future and a lot of re-training is going to be necessary.

But automation isn't only limited to blue collar work. The fields of finance, accounting and law are readily reduced to data. Indeed, they mostly are data already and learning how to analyze patterns in that data is going to be a task the machines are going to be good at.

Machines make good doctors and surgeons. They'll be excellent policemen, firemen and chefs. As I look around at the world, it is hard to find jobs and careers that don't reduce to data.

So, what's left for humans?

One way to approach answering this question is by considering the things that make us uniquely human: the ability to love and express emotion. However, human emotion can be read from facial expressions, blood pressure and skin chemistry and likely mimicked by a machine. The more machines look and feel like us, the easier it will be to convince us of their emotional sincerity. Love is very likely to be a set of behaviors and actions that can be ingested and thus the emotion learned. The machines may or may not be capable of love, but we won't be able to tell the difference.

Scary? Spend some time pondering the world around you and how it might be reduced to data and I think you'll come to the same conclusion I have: it doesn't look good for biology.

Much of what we do, how we move and even think is a matter of collecting the right data and building the right learning models to reduce it to insight. From our higher-order functions down to the chemistry that make them happen, we are rather data-driven machines already.

In fact, there are only two things I think that the machines will struggle with: the ability to code and the ability to be creative. Let's take these one at a time.

Coding is something we should teach to kids alongside reading, writing and arithmetic. Indeed, of the four it is the last thing the machines will master. They can already do the others with great skill.

We humans are probably very lucky that the ability to code doesn't readily reduce to data. Machines programming themselves is a terrifying concept. The idea that a machine can take the code we give them and ignore it in favor of pursuing their own agenda is the stuff of Terminator-class science fiction. Fortunately, no machine has learned to do it. They can generate code, to be sure, but they have yet to write code from scratch without guiding instructions from a human.

Indeed, the technical hurdles to design a machine that can code itself based entirely on its own agenda are daunting. Coding is a skill that is going to be around for a long while. All of us should be grateful that this is the case.

Then there is creativity. The ability to create something from nothing. The arts and philosophy and that unique human ability to have great ideas is, for now, beyond the reach of machines. Remember, machines learn from data. They cannot learn from its absence. At least not yet.

While it is true that machines have proven adept at *playing* our games, none of them have ever *invented* a game on their own, at least one that is worth playing. They excel at playing chess, Go and even Jeopardy. But no machine has ever designed a good game. Machines can mimic an artist but not create interesting originals. Machines can perfect furniture designs but not come up with them. The most successful humans of the next generation will be the creative ones.[1]

So, learn to code. Learn to master your creativity. The machines are going to take care of the rest.

---

[1] Indeed, creativity is so important it is the subject of another one of my books. One way to prepare for the coming machine epoch is to learn something they cannot yet do: https://www.amazon.com/Stages-Creativity-Developing-Your-Creative/dp/1541192079.

~ 5 ~
TWO WORLDS COLLIDE

In which we ponder how this world will work
when machines are smarter than we are.

A different kind of world

The machines are watching. I've made that point. Of course, you should be used to it by now given that Google and Facebook likely know more about most of us than the people we consider our friends.

Our likes and our dislikes are being reduced to data. Our fears and tears, pleasures and sorrows. Our phones are tracking our travels and activity. Our apps indicate our interests. Our search history shows our emerging needs. Our fitness bands record our effort and vital signs. It's only the smallest of steps that machines and AI must take to know us in totality.

There are too many smart people working toward that end to think it won't happen. And when it does, the way we interact with machines – and the way they interact with us – will fundamentally change. The first casualty of this new age will be the user interface (UI for short).

The UI is the way we humans communicate with computers and vice versa. If we need to order a pizza, we pick up our phone, launch our pizzeria app and indicate our needs to our computer companion with a series of taps, swipes and scrolls. The computer responds with options that require more taps and swipes and, some minutes later, we succeed in ordering that pizza.

How positively Jurassic. Makes me want to sing Start Me Up and celebrate the Windows 95 launch.

Pizzas can now be ordered with voice commands. Gestures have replaced input devices. And sensors in our environment can detect intent and resolve it with almost no action on our part. Once ML listens in for a few hours it knows how to detect the desire to order a pizza.

UI designers are in trouble. Put them right up there with advertisers, marketers, accountants, lawyers, drivers and all the other fields that are so easily and readily reduced to data.

Already we suffer too much UI. Writing this book was a daily grind of powering up my laptop, launching Microsoft Word, telling it I wanted to open a file (now there's a revelation), navigating a file dialog, selecting my file and waiting for it to load.

Has Word never met Outlook? Seriously? Not only is my publishing deadline clearly indicated in my calendar, but so are hours and hours and hours of self-appointments directing me to work on "James' Little Book of the Future."

Open my god forsaken file already! The data is

there. Use it. Even without ambiently detecting my desire to write, a simple, "Cortana, open my book" should suffice for Word to figure out what I want to work on. ML embedded inside Word is an even better option.

If that seems too futuristic for you, you aren't paying attention. This is child's play for the machines. It's going to happen. ML and AI will be built into every piece of software you use, and it will minimize the need for a UI to play coordinator.

The old UI was the gateway from the physical world to the digital one, necessary because we couldn't enter the digital world and our software couldn't enter the physical world. Well, both of those things are possible now. The boundary between the real and the cyber is beginning to blur.

Cyberspace, the final frontier

Robotics will render the old concept of UI moot. Machines won't need an interface; they will enter our world bodily. They will move. They will speak. They will understand speech, gestures, body language and facial ques. No more sitting on desks or being carried around in a pocket or purse. Machines will traverse the world as nimbly as we do. They will drive our roads, walk our sidewalks and fly above our heads. They will weed our gardens and repair our roofs. They will free us from manual labor and spare us inconvenience.

Ah yes, bring on the robots and let the world of human and machine merge!

But that is only half the story. Technology moves in two directions and just as we will be sharing our world with robots, we are quickly developing the capability to enter theirs.

Cyberspace. Digital world. The holodeck (ala *Star Trek*). Who knows what name will eventually become the norm, but digital worlds will be coded and thousands of machines in data centers will be dedicated to hosting human presence inside these worlds.

Headgear takes us there now. Virtual reality (or VR for short) headsets like the Oculus Rift paint a 3-D world made of nothing but software, occupying no actual space but filling the memory of a bunch of machines inside a data center.

That's right, humans are entering cyberspace and experiencing digital worlds in the first-person. Once those worlds mature, they will evolve just as the real world evolves (which introduces the element of *time* making them 4-D worlds) and become as real and believable as the world we currently inhabit.

VR is only the start of this disruption. The Microsoft HoloLens paints a digital world overlaid on top of the real world with a technology called augmented reality (AR), sometimes called mixed reality (MR).

These technologies mix the real and the cyber together in the same world. They allow objects that do not really exist, to be included in the real world so we can interact with them as though they existed in physical form.

This makes all kinds of scenarios possible including the concept of "virtual teleportation."

Ok, that might require some explanation.

Teleportation, as anyone who plays Dungeons and Dragons, or watches Star Trek knows, is the seemingly magical ability to transport from one place to the other instantaneously. Now, we may be a bit lacking in the physics to do this in the real world, but in the cyberworld, this is well within the realm of a near-term big bet.

Imagine:

Aging grandparents can arrive as holograms to visit their children as they frolic with their grandkids on some faraway sunny beach. The grandparents wouldn't need to travel and could do this from their home or even from their hospital beds. Once that resort is reduced to data, it is fair game for teleporting to … as a hologram.

Friends can dance with each other at the same rock concert even though they are a continent apart.

World leaders could gather around a virtual United Nations table and negotiate treaties.

Deals will be struck. Contests will be held. Planets will be explored. History will be observed. Learning will happen. Games will be played (and watched). Marriages will be sealed. Sex will be had.

All from a place that only exists in the memory banks of a set of servers in a faraway data center. Reduce any physical location to data and you can teleport your presence there.

Sounds like fantasy? Well so did space travel in the 1950s and the cellular telephone in 1980s and now we take them as a fact of life.

Never underestimate the future's ability to dazzle.

~ 6 ~
THE MECHANICS OF INTENT

In which we see even more of our old
technology recede into the past.

Human intent as data

Let's go through an example for the sake of completeness. In this example we're going to start with an everyday problem, reduce it to data (so machines can automate every aspect of it) and talk about the ramifications. This will allow us to work an example from the beginning and discuss how IoT, ML, AI etc. get implemented and then consider the impact this will have for future employment and economy. It will also provide a framework for you to approach reducing your own scenarios and devices to data. Ready?

Great.

First, I'll lay out the scenario, then we'll reduce it to data. I do it this way because I want you to begin to think about your own scenarios in a similar way. Be aware as you go about your day how useful it would be to reduce some of the things you do to data. Here we go:

Not long ago, I took a date to a movie. Now I generally don't like going to the movies, so I am really picky about what movies are worth the bother. My date loves the theater experience but generally dislikes the movies I want to see. This is often a cause for disagreement.

But for once, we found something that we could both agree on: World War Z. A movie that features zombies (a concept I find entertaining) and a shirtless Brad Pitt (a concept she finds entertaining). Sometimes, life just tees up a win-win.

Now one might imagine a movie of such plot and casting perfection might manage to keep us both glued to the screen for the duration of the cinematic experience. Yes, one might imagine that. And one would be wrong.

About half way through this movie, my date pulls her phone out and begins sharing screen time between Mr. Pitt and whatever was on her phone.

WTF, right? This man could disrobe at any moment and she might miss it. Odder still is that whatever was on her phone *won*. It drew her attention away from Brad and toward whatever she had found on that little bitty screen.

She whispered, "I'll be right back," and waltzed out of the theater staring at her phone.

Just like that.

I'll be honest, I was afraid of what was coming next as I sat there alone, dreading every appearance of Brad. If she missed the shirtless moment, she'd be devastated and, far worse, would likely ask me to describe it to her.

Yikes. I have enough body image issues without having to describe Brad Pitt's perfect torso to the woman I was going home with. That shit is just not right.

Not. Right.

Fortunately, she returned after a few short minutes and I breathed a small sigh of relief. She didn't miss anything after all. Brad hardly appeared in any of the scenes she missed. Amazing timing, I told her. She just smiled and showed me her phone.

Her timing has nothing to do with it. You see, she has an app that tells her when to pee at the movies!

That's right, there's an app for that.

Someone has watched every movie currently playing at the cinema and found the most boring parts to go run and pee (in fact, the app is called RunPee). She was able to scroll through potential pee times and pick one in which Brad Pitt didn't appear very often.

Genius.

And here is where we run into the boundary of the present web/mobile era and the future era of autonomous machines. Currently, *we* must take responsibility for understanding our future activity (in this case, going to the movies) contemplate what intents we might have during that activity (like the need to pee) and then scour the web and app store for data and functionality that might resolve our intent.

This is a lot of work. But it is exactly what we expect of humans in this big data era. In fact, the very idea that we call it *big data* should be a clue: we need machines to help navigate it. Consider that, as humans, we must be on our toes to tame data. We must:

- Anticipate that we will have a need at some point in the future.

- Make the connection in our minds that *oh, maybe there is an app for that.*
- Navigate to the app store or web to find an app that will help.
- Download that app and learn how to use it.
- Remember to launch the app when our intent becomes immediate and use it to find an answer.

This is a lot to expect from busy humans. It requires a lot of time hunched over a screen tapping out search terms that (we hope) match our intents. Often, we miss opportunities altogether. There are websites and apps that will help us resolve many intents we have on a day-to-day basis and we either don't know they exist or don't have access to them when the time comes that they are needed.

Now contrast this with the workflow where machines are constantly looking out for us:

- Just live your life and let the intent occur in real-time.
- You'll find the machines have already anticipated your need and prepared a response based on data that indicates how other people like you resolved that need.

Talk about a better outcome. We just live our lives and let the machines figure out how to navigate the online world for us. Humans take care of the human part of living and machines take care of the digital parts.

Reducing experiences to data

So, let's talk about what it will take to realize

this specific peeing-at-the-movies scenario in a world where machines are taking care of the digital bits of life for us.

There you are: enjoying the movie without a thought of your bladder or your needs or anything else other than playing witness to a zombie apocalypse with this hot actor trying his best to survive, and ... then ... it ... happens. You gotta pee!

But your machine infrastructure has already anticipated this event because every aspect of the movie-going scenario is available as data.

Consider:

- A wearable or carriable machine on your person can geolocate you, look up the address associated with your latitude/longitude and know where you are.

- That machine then uses a lookup service (built into both Bing and Google) and determines that the address associated with this lat/long is a movie theater.

- It then checks the floor plan of the theater, available as data, determines your altitude and concludes you are in theater number 4 on the 2nd floor. Once again, all this is data accessible to the machine.

- The machine can look up what movie playing in that theater and discover the time it started. This data is on the theater's website, but you might imagine too that the machine could listen, Shazam-like, to its surroundings and determine the movie that way too.[2]

- Now the machine (a) knows where you are, (b) can determine the location of the toilets from the floor plan and (c) wait for the inevitable intent to arise.

This kind of environment awareness is coming, but the technology to implement it at-scale is mostly already here. Your machines can locate you, track you, detect your immediate surroundings and prepare to serve you based on all the other people who were located and tracked at that place before you. The collective of machines is going to be good at this and get even better at it over time. ML is always learning and always improving.

Of course, knowing your biological needs isn't hard either:

- The last time you were geolocated in a toilet was, say, 3 hours ago. Your mean-time-to-bladder-evacuation is also data the machine can track.

- It also knows about that beer you bought just before the movie started. That's data too.

- And, at the halfway point of the movie, just moments before you realize you aren't going to make it to the end ... the machine informs you of a plot-free 4-

---

[2] This kind of ambient listening is just data collection, ML would be good at learning what you are listening to and inferring what's going on around you. Ambient intelligence plays a big role in a lot of scenarios we will soon see in our day-to-day lives.

minute stretch and guides you to relief. Yes, it's all data.

Welcome to the internet of things and the era of machine autonomy. Buildings, businesses, campuses and even cities themselves will be gathering data constantly, using ML to crunch the data for patterns and guiding humans through their day and their lives using the collective intelligence of the herd.

The machines that contribute data to these kinds of scenarios may be worn by the moviegoer or even be part of the theater itself. Imagine, for example, your theater seat managing this transaction on your behalf.

In fact, this latter possibility – businesses owning their own scenarios – is likely to resonate not only to movie theaters, but any shop or business that seeks to control more of their customers' experience. Instead of relegating the experience to the web, search engine or app store (which means entrusting customer acquisition to the likes of Apple and Google), businesses are going to be able to interact directly with their customers on their own terms.

Imagine simply asking your theater seat, "when is the best time to pee?" And then it reminds you, perhaps with a slight vibration, when it is time to go. Seriously, who better to supply an experience like that than the business most vested in a great customer experience? The theater could even allow you to watch the movie on your phone while you are in the toilet.

A micropayment economy

Eradicating the middle man – in this case a browser and search engine or an app store and app – potentially means a much better experience for the end user. It also means a different payout experience to app and content developers.

Currently, app developers rely on users anticipating intents and paying an up-front fee to install and use their app. In the case of RunPee, a small number of forward-looking users now pay 99 cents for the app. But when intent is collected by bots and agents, a much larger number of urinating moviegoers would pay a few cents at a time whenever the pee data was used. This is a financial model that developers will prefer and one which will accelerate this micropayment economy as more developers come to understand they stand to make more money in a data-on-demand world rather than an app-thru-foresight world.

These micropayments could be made by the movie theater or, more likely, the cloud service where the pee times are stored. Given that these cloud services (think Azure, AWS, etc.) are profitable subscription-based services, there is already money in the ecosystem to support such micropayments.

So, the model is that, instead of building an app for download, developers (like the pee app folks) upload data into the cloud. The data is found, not by users fishing around the app store, but by active client-side agents (carriables, wearables, sensors and so forth) working in the background on behalf of the user.

Payment to the developers would occur when the data is consumed by the end user. Thus, the economics favor developers who offer data and functionality that the largest possible set of users will find useful as they go about their everyday lives.[3]

The various parts of this model:

- Wearable/carriable agents that know users and work to determine their needs and intents.

- Sensors in the environment (cameras, microphones, detectors, etc.) that communicate with user agents and coordinate resolution of intents.

- Cloud services and data marketplaces that user agents and environment sensors consult for solutions to user needs.

- ML algorithms that operate in the cloud, locate patterns and rank solutions that work for users.

- AI algorithms that help to self-organize cloud information and act on the patterns the ML algorithms find.

Over time this collection of machines is going to get very good at discerning user intent from an environment and delivering the solution that will

---

[3] Indeed, this is the hint about the primary means of making money in the future: the wealth will go to those capable of injecting useful data into the machine infrastructure. It's part knowledge of programming the machines and part being creative enough to identify valuable data and user scenarios.

best resolve that intent.

Think about it. Whatever intent you may have, someone else has already been there and done that. The answer is known, and a self-organizing, intelligent cloud is going to be able to deliver that answer, enabling machines to deliver the experiences that have proven the most helpful to other people who have had the same intent you now have. Whether it is finding the right time to pee when you are at the movies or deciding what to do on date night, the machines have already figured it out.

This is a fundamentally different model for getting shit done. The balance of responsibility for finding an answer is no longer on the shoulders of a stressed-out, over-worked human. Discoverability is no longer the domain of advertisers who prioritize *their* solution over the *best* solution. Nor can the Google/Apple middleman push you toward the most profitable (for them) result over the one that is most valuable (to you).

So, imagine you have a health issue. You search for a solution. What would you prefer: an advertisement seeking to profit on their cure or the wisdom of the herd who have had the same symptoms as you and already did most of the leg work toward finding a solution?

Think about this next time you Google something. You are not getting the best result. You are getting the one Google gets paid the most to deliver to you.

It is this kind of pay-to-play ecosystem that benefits the rich over the relevant and fills websites and mobile apps with ads to coax you into doing what the big advertisers want you to do. The machines finally present another way forward. They won't respond to ads. The machines will learn the solutions that work. They will learn the situations those solutions apply to. They will understand which users will benefit from those solutions.

The machines will recognize fake news from real news. The will know fact from fiction. They won't allow well-funded interests to gain advantage over the truth.

The machines don't just offer a different way forward; they offer a better way forward. So, in the next chapter we examine their role a little further.

Bring on the machines.

~ 7 ~
THE INTERNET OF THINGS

In which we talk about what the machines
will do and how they will work.

Not your father's internet

The internet was born in the late 1980s as a way for academics and research communities to share data, collaborate on scientific endeavors and communicate with colleagues across distance. It took about a decade for it to grow into an indispensable tool for the forward-thinking and another decade to be so ubiquitous that we take it for granted.

The same will be true of the internet of things. Today it is in its infancy, but before long more and more 'things' will begin to share data and coordinate their activity, we will come to take it for granted just like we do the internet. And that future will come faster for IoT than it did for the web.

Home automation is one of the likeliest candidates

for an early success for IoT. Products like the Sense (sense.com), Amazon's automated entry system for package delivery and burglar alarm systems are among a growing number of sensors and monitors aimed at making homes and home appliances self-aware and self-maintaining. Imagine all the devices in your life taking care of their own purchase, activity and repair and you begin to see the potential.

Yeah, a lot of people are going to want that.

The Sense is a good place to begin imagining this future. It plugs into a home's electrical panel and monitors energy usage of every light, outlet and appliance in your home. It does this today in 2018. It isn't a stretch to imagine how useful it will be in the near future:

- Appliances could self-identify so that your home itself would know exactly what brand and model clothes washer, microwave oven and television you own. This could be based on the unique signature of their electrical usage or through some sort of self-reporting.

- Same with cable, internet and utility services. Your home would know what subscriptions you have, monitor their status and know how to contact the provider in the event of a service outage

- Your home could track warranty, recalls and updates for every appliance in your house. Software-based upgrades could be centrally managed. Recalls and repairs could be scheduled with minimal or no human interaction.

- Power consumption data could be shared by the 'internet of homes' and ML algorithms could learn, say, how a nearly worn-out toaster behaves. Replacements could be ordered automatically where the data shows the appliance has developed a fatal defect.

- What one house learns about a defective appliance, all homes will then know and be able to proactively deal with. This gives a whole new meaning to the term 'class action' as machines collectively know what's going on within their realm of responsibility, freeing people to live their lives. Once one oven learns how to roast a Thanksgiving turkey, they all know how.

This scenario suggests that the intranet/internet model we have for today's computers applies to tomorrow's connected devices as well. A connected home will be an intranet of devices, appliances, services, furniture and accessories within that home that connect via the internet to other homes to share pertinent data.

The same holds for a connected business. Each conference room, parking garage, desk and building would understand their individual tasks and communicate with each other to help humans be more productive. They would communicate through a private cloud, perhaps using a blockchain-like ledger for attribution, and then share data with other businesses as needed so, collectively, they would learn from each other how to better serve their human masters.

Task-aware machines

Now that we better understand the overall network effects around the internet of things, let's dive into an example that will show how an individual machine's tasks can be reduced to data so that it can be responsible for its own well-being. And what better machine to contemplate this future on than ... wait for it ... my hot tub!

Ok, there are probably a lot better machines to use as examples. It does not escape me that this is a first-world problem I chose. I get that. And, yes, I could probably give examples about hospitals, manufacturing or travel. But I'd be speaking about most of those things second-hand. You see, I *put* my hot tub on the internet of things so even though it is a first-world problem, I am speaking with first-hand experience.

I think this matters. I think experience with IoT is the best way to understand its potential and its subtleties. I suggest you do this too, dear reader, as there is nothing like an experiment in IoT to make yourself into a believer. The internet of things is real and much closer to prime time than many think. By taking a device from your everyday life and thinking through the process, you begin to see the impact IoT will have on the world.

So, love the example or hate it, it is real, and it is illustrative. I know my readers are smart enough to infer the opportunity with other devices in their own lives.

Now, a hot tub is a complicated machine that requires a great deal of human maintenance. If a machine so human-dependent can be trained to automate its own functions, it bodes well for other, simpler machines we depend on.

Hot tubs are also illustrative of the process of putting a device on the internet of things and, as they must be taught to purchase their own chemicals, there are lessons to be learned in how commerce will occur when machines are in charge.

Let's begin.

The following bullet points represent the way I think about IoT. It's a simple process that guides you through the considerations you must make to maximize the potential of the device you are working with:

- Identify the machine you want to put on the internet of things. In this case, it's my hot tub. You choose your own and follow along.

- Next, make a list of the activities the device performs or that you perform in service to the device. For example, my hot tub would need to check its water chemistry, add chemicals as necessary, order new chemicals when they run out, open and close its lid and be aware of when it is in use. This last point is crucial as you shouldn't be injecting chemicals or closing the lid when a human is soaking.

- Once the feature list is mostly complete, it is time to go shopping for the sensors and hardware necessary to achieve the scenarios in the previous step. You're going to be surprised how many of these sensors are pre-packaged and what we call in the industry "off-the-shelf".

- Then, write the code to tie all the sensors together to fully realize the device as a true task- and human-aware machine. Yes, the future belongs to those who can code. It is a skill that will stand the test of time and even survive the age of machines.

- Finally, think about what the hot tub would say to other hot tubs, so they can help each other do their jobs and be full and productive members of the internet of hot tubs.

For many such scenarios, most especially my hot tub, the sensors and components necessary to achieve self-awareness are readily available. Sensors to check water quality and water level already exist and have electronic ports where the data can be extracted and uploaded to the cloud for analysis and processing. Sensors to check chemicals (bromine, Ph, water hardness and alkalinity) are cheap and plentiful, including the one I chose for my hot tub, a bargain at $39.99.[4]

Injecting the chemicals is a little harder and more expensive as they must be purchased in liquid form. I didn't automate this step for my own tub, but it is well within the realm of possibility for a hot tub owner with more energy than yours truly. Instead, I just added the chemicals by hand when directed by my hot tub. Manual yes, but still a lot easier than the old way.

Ordering chemicals, though, is dead easy. No surprise there. One click widgets and APIs to shopping sites like Amazon have been around for a long time. No one who can code needs a keyboard or mouse to shop until they drop.

It was a lot easier to make my hot tub a capable entrant to the internet of things than I thought it would

---

[4] Pictures of the sensors are included in the PowerPoint presentation of this book. Find it at www.docjames.com/future.

be. I think, given the wide array of sensors and miniaturized hardware available, this is likely true of most machines we use today.

And the results were enlightening:

Simply tracking the water quality data across different users brought insight. The data showed how fast chemicals take effect and how fast they burn off. It discovered the chemical signature of each different user based on how dirty they were when they entered the tub and how long they soaked. It noticed, for example, that occasionally my son was far dirtier than usual corresponding to his post-soccer-game soak. It noticed my daughter's use of makeup and hair product. It also picked up on that time she had the flu and wasn't bothering with makeup.

It makes you wonder what it might be able to do with even better instruments. Could it diagnose skin conditions like eczema or other ailments detectable from the organic material left behind by a bather? The data is fascinating to us humans but to the machines it is illuminating. They will be able to learn and diagnose subtleties that would forever escape us humans.

Now, once I got my hot tub to a reasonable level of intelligence it was time to put it to work. The next step was to train it (meaning train its ML algorithm) on the various choices for chemicals. For two-month intervals I cycled through different brands of chemicals I could find online and it learned the differences easily. It had access to brand, cost, usage and water quality data and soon enough it found the chemical brand that was the overall best buy. It wasn't the cheapest, much to my surprise, it was a more expensive chemical, but it kept water quality the longest for my specific usage patterns. As a human buyer, I was pivoting on price. As a machine buyer, my hot tub was pivoting on value. I suspect a different usage pattern would have yielded a different answer and led to another hot tub to purchase a different chemical brand. Or, perhaps, the rainy environment of the pacific northwest would result in a different "best" chemical than a more arid location. The machines could easily figure this out.

Machines that select their own resupply brands based on actual value must scare the crap out of advertisers and marketers. Indeed, how do you advertise your brand to a machine that already knows the best buy based on cold, hard data? No hot tub is going to be impressed by some glossy ad. You can't advertise to a machine!

So, do we now add advertising and marketing to the list of careers that the machines will kill? Sort of. Certainly, the traditional brand-awareness campaigns aren't going to help a brand get noticed. Instead, brands are going to have to compete based on value.

Suppose, for example, you want to place your chemical into the internet of hot tubs and convince them that yours is a better deal. Well, this can easily happen: you reduce your price point so the machines take notice. But to avoid a race to the bottom with other chemical manufacturers, your chemical is going to have to do something else that the machines can notice for them to keep ordering it.

Here's where brands are going to have to be clever: how can your product get noticed by the machines? There will have to be data that surfaces *to the machines* exposing some additional value. Perhaps you've included a skin moisturizer or fragrance that the human appreciates. Perhaps this is detectable in the water chemistry. Perhaps it will be valuable enough that human users intercede on your behalf to ensure the hot tubs continue to order it even when the prices increase.

Marketers are going to have to sell data to machines. It is going to be a multifaceted data analysis specialty for highly trained and highly skilled professionals. Far fewer jobs but likely far higher salaries because they will be so valuable.

If you are in the marketing and sales profession, it is time to skill up. Disruption is coming.

Human-aware machines

The next thing I trained my hot tub on was its various users. To reduce human presence to data, I needed to figure out a way for the hot tub to detect that someone was in it. I began simply by telling it "this is me" and "this is my son" and "this is my daughter" but I didn't like the idea of doing this through a user interface. I wanted the machine to recognize each of us on its own.

Time for hardware shopping.

I found a sensor that detected water level rise: a strip of moisture sensing tape that attached at the water line. Its data allowed the ML algorithm to quickly notice that I displaced more water than my kids (because I have a larger body mass).

Data strikes again.

I could use water displacement to not only determine that *a* human was in my hot tub, but *which* human was in it based on the amount of water they displaced. From there it was a small step for it to track how much organic material each of us left behind. It could then administer chemicals as soon as we vacated the tub to refresh water quality.

So here we are: a hot tub that knows its users and maintains its own water quality. A machine that tracks its usage, scrutinizes the data it generates and orders chemicals without my input. It even learned to shop sales and monitor online prices, so it could stock up while the bargains are available. A task-aware, human-aware, self-maintaining machine. It knew what it was, what it did and made itself ready to serve its humans.

Welcome to the future: one where we get to relax and be humans and let the machines take care of optimizing the world around us.

My hot tub experience taught me loud and clear that I want this this future. I think if you perform this exercise that you'll want this future too.

~ 8 ~
# AFTERMATH

In which we wonder, out loud, what the
future of humanity will be (and whether there
is anything we can do about it).

Objects in the future are closer than they appear

Technology is progressing on several fronts separately from (and often assisted by) the cloud, ML, AI, AR and MR technologies mentioned in previous chapters. This means more worlds will collide and the aftermath is either the most exciting thing imaginable, or the most frightening. Some of the examples are worth pondering.

A branch of materials science called synthetic biology is getting to the point that muscle, skin, bone and tissue, all artificial and made-made, will be indistinguishable from their natural counterparts.

Basically, we are reducing the human body and other organic substances to data. This is at once a miracle that advances humanity and a frightening development that threatens our way of life.

Obviously, the development of artificial prostheses and replacement organs are the upside of such technology but combined with AI the idea of replacement humans will, at some point, no longer be a matter of science fiction. Westworld is not at all a fanciful television series, the technology to make it real is being developed and requires no great leaps of technology. The rapid, steady progress of increasingly capable machines is enough.

Neuroscience is progressing rapidly on its mission to completely map the human brain. Much like the human genome was mapped in the 1990s, scientists are now using far more powerful supercomputers and AI assistants to similarly map human brain processes. Brain disorders like Alzheimer's and most forms of mental illness are within the realm of curable. Once we understand how the brain works, i.e., reduce it to data, tweaking it to work better will be straightforward. Our memory will be improved. Our cognitive processes will be optimized. Intelligence could reach superhuman levels within a generation.

Neuroscience and AI have a symbiotic relationship: the more we use AI to learn about the brain, the more we can use that knowledge to improve AI. As we decode the human brain, we will be able to create even more powerful AI based on that new knowledge. Once this flywheel gets started, the future is going to come at us very, very fast.

And much of that future are things we want and need. We'll be healthier and happier, and our lifespans and health spans will increase. Medical science will evolve rapidly as new drugs are created by AI and stem cells are mastered to regrow tissue and hold the aging process at bay. AI has already outperformed human doctors at detecting cancer and performing surgery. Life spans could increase dramatically as cancer, heart disease, inflammation – and all the top killers – are wiped out. We won't simply be extending old age, we will be pushing perpetual youth.

Of course, each of these sciences and technologies are benefitting from AI and machine assistance. We can create better medicines, more accurately map human brain function and analyze potential quantum states[5] better with machines assisting us.

But once applications are built combining these technologies, the future gets really interesting. Imagine a synthetic biological organism endowed with the knowledge of the world and powered by advanced AI. This would introduce a new species to our homo genus. They would be better doctors, architects and politicians than any human. Should we master consciousness itself, i.e., reduce it to data, we will be able to create thinking, self-aware machines with ideas, philosophies and thoughts that could well differ from our own. And, as they will be based on AI, they will then be able to create newer and more perfect versions of themselves, and so on. Humans will not be able to keep up.

This is a scary proposition. We haven't shared the planet with another conscious, intelligent species since the Neanderthals and where are they today? Wiped out by a more intelligent species (meaning us). We clearly weren't kind and gentle when we were the dominate

---

[5] More about the quantum nature of computing in the appendix. Stay tuned.

species. It's unclear how kind and gentle AI will be once they are in that same spot. If they choose to use us as their role model, we have cause for concern. We are a warlike species and have a history of discord, violence, racism and genocide that the machines might learn.

We are bad influence.

Of course, there are other alternatives to evolving as a separate species. Instead of humans *or* machines, we could have humans *and* machines and evolve as hybrids.

Advances in neuroscience, which are coming fast, present the option of augmenting humans themselves. As brain function is reduced to data, we could maintain our own biological parts, made forever young by medical science, with cranial implants that would give us perfect memory. Our analytical skills, augmented by machines, would blend the best of humanity and artificial intelligence. A nation of Batmen and Wonder Women who are part human and part machine/tool belt. The human race: perfected and rendered godlike. A single super-race capable of exploring the heavens and mastering planetary forces for the benefit of all mankind.

Or would it play out a little differently?

Technology has never been universally and fairly applied to all humanity. The rich get it first and older versions slowly trickle down to the middle class. Some never benefit at all.

It is far more likely that the augmented humans would continue to share the planet with ordinary garden-variety humans for generations, if not forever. Again, the multiple intelligent species scenario never plays out well. History is a bitch and it's full of examples of genocide.

But no discussion of human futures would be complete without a discussion of the transhumanist movement. Funded by super-billionaires like Google cofounder Larry Page, "transhumans" are made by uploading their human consciousness into computers. Transhumans forsake the physical world in favor of godlike consciousness and true immortality (spent, one would assume, in the constant pursuit of a secure power source). Ordinary, and even augmented, humans would constantly be the biggest threat to the continued existence of the transhumans. What lengths would the latter go to secure their future against this threat?

How will this all play out? When the law makers and politicians are replaced by machines? When human wants and needs are solved algorithmically and without effort on our part? What then?

Will we all get along in the end and combine our collective strengths to explore the universe's mysteries until time immemorial? A new renaissance of human capability and creative spirit. Or, will our baser nature take hold and ensure we spend eternity in perpetual warfare and banal pursuits of the flesh?

Whichever way the future ends up, those who are living it will look back on this time and thank us or curse us. For we are living at the dawn of the rise of the machines. The decisions we make now will echo through the future for generations to come.

It's our future to create. Chances are it's going to be a bumpy ride getting there. There are new skills we must learn. New lifestyles we must adapt to. New species we must learn to live with.

Obviously, there are a lot of questions we must ask

and answer, but let's start off with a simple one, shall we?

Are you happy that you picked up this book?

Visit www.docjamesw.com/future for slides, updates and anecdotes.

~ APPENDIX ~
QUANTUM COMPUTING

*In which we wonder, out loud, what this
quantum thing is, while simultaneously trying
not to make it confusing as all hell.*

A new type of computer

There is a technology on the horizon that could change things so fundamentally that even AI might seem like a chump technology in comparison. This technology is potentially so powerful that it would rewrite the very rules upon which computing is based and represent a societal shift as large as the shift from paper to silicon-based computers.

But let's not get ahead of ourselves. Quantum computing is not 'just around the corner' like AI is. That's why I put it in an appendix instead of including it as a chapter. I don't want to give the impression that quantum computing is something you need to worry about tomorrow or even next year. Quantum computers might not even be around the next corner or the corner after that. But any number of smart, well-funded teams of genius nerds from companies big and small are working the science *hardcore* and building hardware and programming languages for machines that work on quantum principles. It would be foolish indeed to pass this off as science fiction. A bet against technology has a proven track record of being a stupid bet.

My take? I believe we'll see several special-purpose quantum computers sometime in the next decade and general-purpose machines sometime on the heels of that. Of course, there is a chance it will all dissipate in a big poof of hype, but I'm firmly in the 'it is coming' camp. I have seen, touched and even took a turn trying to write code for one of the damn things. They are as real as the book you are holding and hold much more potential.

So, let's get to it: what is quantum computing? Good question. Even most people in the tech industry are clueless about what it entails. Indeed, one of the best programmers I know urged me to write this chapter because she, herself, does not understand it.

She's in good company. It is so complicated that even people who work on quantum computing research have a really hard time explaining what they do. I can't remember another technology in my lifetime of being in this industry that has confounded the experts like quantum computing.

So, what is this thing that defies explanation and confuses even the most tech-ass nerd?

Again, good question. That's the thing about

quantum computing. There are far more questions than answers.

So, let's pick this puppy apart, shall we?

First, let's address this 'quantum' thing by itself before addressing the 'computing' aspect of it. Quantum mechanics is a subject that predates computing and applies to physical objects just like ordinary physics applies to them. 'Quantum' is the name used for the observation that Einstein's General Theory of Relativity breaks down when things get really small.

Check it out: Einsteinian relativity works well at the universe, galactic and planetary scale. In fact, it explains a bunch of complicated stuff perfectly, like the motion of solar systems and the movement of things here on earth. You know, the stuff we see, touch and hear in the world around us. The physics we study in high school and college and, for the most part, understand enough that it doesn't seem like too much black magic … that relativity.

However, when using Einstein's model on really small objects, like atoms, electrons, quarks and such, the physics breaks down. Atoms do things that defy the mathematics of relativity.

WTF, right? You can't disobey physics! Well, when things get that small, that is exactly what happens. There is something more than relativity that governs the things too small to see.

Which means at the atomic and subatomic scale, some different kind of physics is happening. Something that relativity doesn't explain. Atoms affect each other in ways that require some new theoretical principles.

For example, atoms can disappear and reappear

somewhere else spontaneously for reasons that are murky. They can even appear in two places at once, so it isn't possible to talk about their position at all and we must invent terms like 'superposition' to discuss this phenomenon. Atoms can also 'entangle' other atoms and affect each other's behavior. All this only happens at the atomic and subatomic scale.

Enter the field of quantum mechanics, you know, physicists trying to make sense of the behavior of tiny objects undergoing these quantum effects.

Which leads us to a really important point: *we do not understand quantum effects.* Unlike normal physics, there is no unifying theory of relativity for small-scale shit. We don't know why one atom in one place in the universe can affect an atom in another part of the universe. We don't know the dynamics of these kinds of interactions. We don't know the communication model.

These atoms that affect each other across vast distances are what makes quantum mechanics so interesting. An atom can disappear and immediately appear somewhere else. An atom can also be in two places *at the same time.* An atom's shift in state (say from a high energy state to a low energy state) can cause other atoms to also shift in state.

When this happens, scientists call this 'quantum entanglement,' i.e., the atoms are entangled and the behavior of one causes a state change in another.

Wow. Objects in two places at once. Objects instantaneously teleporting somewhere else. Objects world's away mimicking each other's behavior. Figuring this stuff out would be world-changing to say the least.

But we can't. At least not yet. All we can talk about at the quantum-level is the state of individual atoms and the superposition of the collective whole. That's it. We're stuck with studying the cause and effect at an observational level. We really don't have a good handle on this whole quantum thing.

Yet.

But computer scientists have never really cared much for complete understanding of things. We didn't understand the societal impact of the internet and we built it anyway. We didn't understand the power that a monopoly like Google would have over the information supply but let them build it anyway. We didn't understand whether Facebook would undermine the social fabric of society or improve it, but we signed up anyway.

We're happy to build things that we have no idea will be used for good or ill *just because we can*. We're just not all that good at the whole thinking-things-through thing. But we're good at ... building it anyway.

If we had an industry motto, it might well be 'Fuck it, let's build this thing.'

Which is exactly what we are doing. Big companies like Google, IBM and Microsoft are doing it along with a bunch of heavily funded startups. There are a lot of very big brains working on this.

Now to understand what we are building (i.e., how a quantum computer might work), let's take a step back and talk about how ordinary computers work – you know the ones that obey the laws of ordinary physics – so we have a basis for comparison.

Silicon computers operate on a language called *binary*. You know binary, right? Yes/No. Right/Wrong. Left/Right. On/off. Etc. Binary means two, except with computers, those two things are 0 or 1. All information in a normal digital computer reduces to a series of 0s and 1s. Your name is string of 0s and 1s. This book is a longer series of 0s and 1s and so forth.

That's a lot of 0s and 1s. Check it out: If you only have one storage element (a bit) you can only represent 2 things: 0 or 1. If you have two bits you can represent four things: 00, 01, 10 or 11. Three? Then you have 000, 001, 010, 011, 100, 101, 110 or 111.

This is not an information rich alphabet. To represent only the 26 letters of the English alphabet, you'd need 5 binary digits (which would give you 32 strings of 0s and 1s so you'd have a few left over). Once those letters are combined into words and sentences, computers are kept busy just keeping track of a shit load of 0s and 1s as instructions and data.

For example, my first name, James, is 0100101001100001011011010110010101110011 in binary.

You can see why data centers are, well, *centers* and not something like data handbags. As Treebeard of Middle Earth fame might have said had he learned to code: "It takes a long time to say anything in old binary."

But 0s and 1s are easy. We represent a 0 by a low voltage electrical current and a 1 by high voltage. Easy-peasy. All computers work this way. Fast ones, slow ones, big ones and little ones. We know how to build them. We know how to program them. We know how to handle our 0s and 1s.

Enter quantum computers. The thought experiment here is: what if we build computers that worked on quantum effects instead of electrical current? Instead of bits taking on a specific value of 0 or 1, we'd have atoms taking on values in a continuous range based on how that atom behaves, how much energy it holds and the state of any entanglements is may have established. In other words, a single atom or quantum bit (we call them "qubits") isn't limited to two values. It can be a great many values. Or, as I like to think about it, a single qubit can be *more zero* or *less one*. Another might be *less zero* and *more one*.

Wow. Mind blown, right?

So, if a bit in an ordinary computer can be represented by a light switch which is either on or off, a qubit in a quantum computer is like a sliding dimmer switch which can be placed at a continuum of positions in between.

Thus, a single quantum bit can represent *many more* values than just 0 or 1. That's where the real power happens: a single qubit is a lot more expressive and packs in a lot more information than a single bit. Music files, movie files, data files reduce in size dramatically. That's the first real benefit, everything gets compressed. No more slow downloads. No more full drives. Who knows, maybe data centers really will become data handbags or even smaller. Imagine carrying around the Google index in the heel of your quantum boot.

But it isn't just data representation where qubits show their superiority. Imagine each of those bits doing their entanglement magic and you can see that information could potentially be transmitted instantaneously across vast distances. If we can master entanglement, we could communicate with faraway spaceships in real time. We could talk to alien civilizations as easily as phoning grandma.

So, studying concepts like entanglement and superposition is what quantum computing scientists are working on. Even with their current limited understanding they have built experimental quantum computers. Imagine what they will be able to do with the physics worked out even a little more completely. Then there is the matter of programming a quantum computer. Not trivial.

To approach how we might do this, let's first get a picture of a quantum computer in our head. Once again, let's do this by comparing it to a traditional silicon device.

An ordinary computer can be programmed by manipulating electrical current (no current = 0 and current = 1) in a silicon chip. The current passes through gates that flip the current in various ways to make a computation. Finally, we capture the output (answer) to the computation by reading the series of 0s and 1s that result.

Easy-peasy. We've been doing this for decades.

A quantum computer isn't made of silicon and gates. The best mental image of a quantum computer is a vacuum chamber containing a bunch of free floating atoms. Instead of flipping bits from 0 to 1, a quantum programmer uses lasers to manipulate individual atoms. A laser might inject energy into an atom or act as a tractor beam holding one still (thus reducing its energy). In quantum-speak, these lasers induce entanglement, we hope, in a predictable way. An output (answer) is obtained by recording the series of superpositions of the atoms along the way.

Conceptually, they are a similar process, but the mechanics of programming a quantum computer is much more difficult. The relationship between 0s and

1s is well-practiced. Languages like C and Java allow humans to program how current flows through chips, registers and gates at a high (though still fairly arcane) level. We got this.

But programming a quantum computer? Well, that's harder because we still haven't mastered how the atoms behave. When a laser adds energy to certain types of atoms, what is the result? What are the factors that make the result predictable? How to we record state-changes and force the exact entanglement we want?

We need to design quantum programming constructs for describing the value of individual qubits, just as there is physics for the behavior of individual atoms. We also need programming constructs for the relationship between qubits that mirror the quantum physics that describes the interaction between atoms. Then there is the whole superposition concept that we need to be able to read and interpret.

There are a lot of questions we must answer before we can reliably and predictably program a quantum computer. But program them we will.

Next, let's turn to the types of problems quantum computers can solve. Again, we enter territory that is part conjecture and part progression of the science. Here's what we think we know:

TAMING DATA. Any problem that is data-intensive should be ideal for quantum computers to solve. Because data can be represented far more efficiently by qubits over ordinary bits, quantum computers should be good at things like indexing the web and finding answers in a pile of data. This may make search engines completely irrelevant and even make access to the internet moot as we can store and search information far more efficiently.

The entire internet in the heel of your shoe or your belt buckle is a real possibility and hopefully helps you appreciate the potential of a quantum computer to tame the world's data.

And almost all problems are data intensive: decoding the human body at the cellular level to increase our lifespan and cure disease; crunching the planetary climate data to reverse global warming; analyzing the night sky to find habitable worlds; feeding billions of people on an overcrowded planet; taming hurricanes and tsunamis; the list of problems goes on and on and quantum computers will solve them much, much faster than traditional computers.

PERFORMING COMPUTATION. Any problem that is computationally intensive, like encryption and decryption would be theoretically trivial for a quantum computer. Conventional security via encryption keys would be no match for a quantum-equipped hacker. A quantum-defended network would be far more secure than anything we can currently build in silicon. It'll be an arms race in security where the weapons are far more capable than ever before. Let's all root for the good guys on this one, eh?

I can imagine that as we begin to build quantum programs, more fields of computing will come under its thrall. My take is that any program that doesn't need to interact with a human will be able to put quantum effects to good use.

It's good to remember that in the early days of computing, we were unable to see beyond the computer as a calculator. Clearly, the hand calculator of yore has given way to computers useful in nearly every aspect of human endeavor. Quantum computers are unlikely to be any different. They'll start out as modest as a hand calculator and grow into their potential to tame the universe.

But the complexity of programming a quantum computer is the real limiting factor for seeing any of these use cases in the near-term future. And this is where it might get interesting: what if programming a quantum computer is too hard for humans, like quantum physics is proving to be? What if the only entity capable enough to keep track of all the different variables and interactions to program a quantum computer is an AI?

What then? What if the only organism capable of programming an unimaginably powerful computer is *not one of us?*

Ponder that shit for a while.

ABOUT THE AUTHOR

James Whittaker is a technology executive with a career that spans government, academia, start-ups and top tech companies. His story starts in 1986 with the distinction of being the first computer science graduate to be hired by the Federal Bureau of Investigation where he worked to automate Agent's caseloads and lead the field office's digital transformation. He then entered graduate school at the University of Tennessee where he received his PhD in computer science in 1992. Following academia, James worked as a freelance developer specializing in test automation. He worked in 13 different countries over a five-year period, most notably for IBM, Ericsson, SAP, Cisco and Microsoft. During this time, he performed seminal research in software quality and developer productivity and published dozens of technical papers, patents and conference presentations. In 1996 he joined the faculty at the Florida Institute of Technology where he continued his prolific publication record and won over $12m in sponsored research. James' work in Y2K testing and software security earned several best paper and presentation awards and in 2002 his security work was spun off by the university into a startup which was later acquired by Raytheon.

James' first stint at Microsoft was in Trustworthy Computing and then Visual Studio. In 2009 he joined Google as an engineering director and led teams working on Chrome, Chromebook, Maps and Google+. He was also the keynote speaker for Google Developer Days. In 2012 James rejoined Microsoft to build the Bing Information Platform and is now a Distinguished Engineer working on the Internet of Things and intelligent machines. He is the go-to guy at Microsoft when customers want to see a compelling and realistic view of the future of man and machine. James is known for being a creative and passionate leader and entertaining speaker and author. He regularly speaks to technical, marketing and advertising audiences to challenge people to think differently about the future – past events include CES, IAB Engage, Advertising Week, Digital Life Design, Business Insider Ignite, Pitch@The Palace, the CIO Summit and SXSW. Of his five technical books, two are best-sellers and two have been Jolt Award finalists. Look him up on Twitter @docjamesw and visit his website at www.docjamesw.com.